程安琪 著

# 老火湯

# Wealth of knowledge in A Bowl of Soup

養生保健——從一碗湯開始

## Nourishing your health with soup

# 目錄
## Contents

# 煲湯食材的份量、水量、時間和火候要如何拿捏？

## 煲湯要放多少水？

一碗湯約 220 毫升，一般一家六口，每人兩碗，基本水量就是 2640 毫升。煲煮老火湯通常約為 3 小時，按煲 1 小時水分蒸發流失 10% 計算，煲 3 個小時要額外增加 30% 的水。

基本水量：220 毫升 × （2 碗 × 6 人）=2640 毫升
煲 3 小時的水量 ：2640 毫升 × 1.3 =3432 毫升（約 3.5 公升）

注意：

1. 水量的計算也要考慮到食材原料的份量與特性，如豆類、米糧、乾貨或藥材等容易吸水，煲湯的水不妨多加一點；若使用容易出水的材料，如蔬菜、瓜果類等材料，煲湯的水量可以減少一點。
2. 煲湯的火候也會影響水分蒸發的速度，火越大，蒸發得越快；火越小，蒸發得比較慢。
3. 隔水燉的湯，由於水分不會蒸發掉，因此水量無需額外增加。

## How much water is used for boiling soups?

A bowl of soup is about 220ml, for a family of six, you will need to cook about 2640 ml of soup if each person takes two bowls. It usually takes about 3 hours to boil the soup, if 10% of the water is evaporated in an hour, you will need to add an extra 30% of water to make up for the loss.

Basic amount of water: 220ml x(2 bowls x 6 persons)=2640ml
Amount of water to cook for 3 hours:2640ml x 1.3 =3432ml (about 3.5 litre)

Note:

1. Amount of water used also depends on the amount of ingredients and their characteristics, for example beans, grains, dried ingredients or herbs absorb more water as they cook, thus it is necessary to increase the amount of water used for making the soup. On the other hand, vegetables and gourds give out water thus less water is needed.
2. If the soup is cooked over a higher heat, it will cause water to evaporate faster. The amount of evaporation is reduced if the soup is simmered over low heat.
3. For double-boiled soups, there is no loss of water, thus no extra water is needed.

## 如何拿捏材料的份量？

煲湯主要材料份量的拿捏，以每人所需份量，乘以食用者人數最為理想。以下為一般份量計算方式：

肉類、海鮮類平均每份 100 克
蔬果類平均每份 150 克
糧食類平均每份 50 克

注意：

1. 要如何準確地衡量材料的份量，這就必須靠日常不斷煲湯的經驗和家人的口味自行揣摩，有

# Boiling Soup: how much ingredients and water? Cooking time and heat control.

## 如何拿捏煲湯的時間？

老火湯一般是將鍋裡的食材煮沸後，再用文火（中小火）慢慢煲上兩三個小時，由於火候足，湯味濃郁、鮮甜。

燉湯一般是將食材和藥材隔水猛火燉，起碼要燉3個小時以上，而且蓋上密封，連湯的香氣也保留在內；因為湯水彙聚食材和藥材的精華，滋補效用更明顯、更具香氣。

注意：煲不同食材的時間相對有異，煲魚湯大約1-2小時；肉湯約2-3小時；煲蔬果類大約1-2小時；燉湯則需要增加一些時間，約3-5個小時。

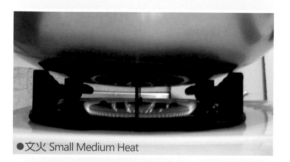

●文火 Small Medium Heat

●武火 High Heat

## What is the cooking time?

Slow boiled soup is cooking the soup over small-medium heat for 2-3 hours after the soup has been brought to boil. Soup boiled in this manner is rich and full of flavour.

Double boiling is to sit a jar of soup in another pot of water over high heat and slowly boil it for at least 3 hours. As it is tightly covered, the aroma of the soup, and the essence of the ingredients and herbs are concentrated in the jar, giving you a soup that is highly nutritious and full of flavour.

Note:
Different types of ingredients require different cooking times. Fish soup takes about 1-2 hours, meat soup takes about 2-3 hours, vegetables soup takes about 1-2 hours. Double boiling takes a longer cooking time, about 3-5 hours.

---

些人喜歡重口味，有些人喜歡清淡些！

2. 煲湯食材與水的建議比例：原料與水量按1：1.5，煲出來的湯色澤好、味道好，而且營養價值高。

## How much ingredients to put?

The amount of ingredients needed depends on the amount each person takes multiplied by the total number of people. Here is a basic guideline:

Each serving of meat and seafood is about 100g.
Each serving of vegetables is about 150g.
Each serving of grains and cereal is about 50g.

Note:
1. To figure out the correct amount of ingredients used, you need to know your family members' tastes and needs and cook accordingly.
2. Using the ratio of 1:1.5 for ingredients and water will yield a nutritious soup with a good colour and flavour.

# 煲湯 9 不要訣
## Soup Boiling Tips

### 1 肉類不要直接下鍋
### Do not add meat directly into the soup

煲湯的肉類不要因為省麻煩，直接丟下鍋煲煮。肉類一定要先汆燙，去血水和多餘的脂肪，避免湯水渾濁和帶有腥味。

It is important to scald meat in boiling water first before cooking it in soups. This is to remove blood water and excess grease so that the soup will not turn out cloudy with an unpleasant smell.

### 2 調料不要放太多
### Do not add too many ingredients and seasonings

有些人以為，調料放得越多，湯的口感就會越好。其實，這種想法和做法都是錯誤的，如果調料放得過多過雜的話，湯可能就會串味，反而會影響湯的口感。

It is a misconception that the more ingredients and seasonings you add to the soup, the more tasty it is. Adding too many ingredients will over-complicate the flavour combinations and affect the final taste.

### 3 藥材不要亂搭
### Be careful when using Chinese herbs in soups

有些人想通過喝藥材湯給自己進補，卻又不知道藥材的特性，比如寒、熱、溫、涼等。因此，煲湯時必須根據自己所需來搭配添加的藥材，如寒氣過剩的人最好選擇溫性的中藥材；體質熱的人如果吃溫性中藥材的話，很容易出現上火的現象。

Soups brewed with Chinese herbs have many health benefits, however it is important to understand the characteristics and nature of each individual herb and choose those that are suitable for one's body constitution. Choose herbs with a warm nature if you have a cold body constitution. Refrain from using herbs with a warm nature if you belong to a hot constitutional type as these will cause heatiness in your body.

### 4 中途不要加冷水
### Do not add cold water in the process of cooking

煲湯中途不宜加冷水，因為湯料表層蛋白質在高溫情況下會突然遇冷收縮，影響營養和鮮味的釋放。加水最好加熱水，而且分多次少量添加，這樣對湯的風味和口感影響會小一些。

Adding cold water in the process of cooking will affect the release of nutrients and flavours from the ingredients thereby affecting the taste of the soup. It is advisable to add hot water a little at a time if you need to top up the water.

### 5 加鹽不要過早
### Add salt at the end

煲湯時，如果鹽加得過早，肉裡面的蛋白質就會出現凝固的現象，不容易溶解出來，還會使湯的顏色變暗。最好在出鍋之前才加鹽，這樣不僅會保持肉質的鮮嫩，還會增加湯的口感。

When cooking soups, adding seasoning at the end is the best way to go. When salt is added at the beginning, the protein in meat will coagulate and the soup will turn dark in colour. Add salt last to keep meat tender and the soup vibrant and tasty.

### 6 不要一直用大火
### Control heat properly

煲湯時，先用大火把湯煮滾，然後改成小火慢煲。因為如果一直用大火煲煮，肉中的水分很容易流失，而且很容易乾鍋，非常影響湯的口感。

Bring the soup to a boil over high heat, then simmer over low heat to finish cooking. Cooking it over high heat throughout will dry up the soup and lose the depth of flavour found in slow cooked soups.

### 7 時間不要太長
### Do not overcook

許多人以為煲湯時間越長，營養越豐富，其實這是大錯特錯的。煲的時間太長，營養會流失，而且導致痛風的嘌呤含量會增高，適得其反，破壞湯的營養，影響健康。

It is not true that the longer the soup is cooked, the more nutritious it is. When soup is cooked for too long, nutrients will be lost and the level of purine will increase. Purine is related to the development of grout.

### 8 泡發海味、乾貨不要偷懶
### Rehydrate dried seafood and other dried food ingredients well

煲湯的海味和乾貨必須先用清水沖洗，再用清水泡軟後才入鍋煲煮，泡發的乾貨或海味讓湯的味道更濃厚、更具獨特的香氣。

It is important to rehydrate dried food ingredients well before cooking. First rinse well, then soak in water until soften before cooking in soups. These rehydrated ingredients will give a soup that is rich and packed with flavour.

### 9 豆類、穀類不要直接下鍋
### Soak beans first before cooking

豆類、穀類要事先浸泡，吸收水分，煲煮時較快煮軟。此外，沒有經過浸泡過程的豆類，先炒過再入湯，可以達到相同的效果，味道也會更香。

Soaking beans or grains in water allow them to absorb moisture so that they cook faster in soups. Alternatively, you can fry them beforehand for the same effect, this also enhances the flavour of the soup.

# 吃得好，也要吃得對！
## Eat Well, Eat Right!

### 豬肉
### Pork

▶保健功效 Health Benefits

具有滋陰潤燥、補腎養血、益氣強身的作用。

Replenish the Yin, moisten dryness, nourish the kidneys and blood, replenish vital energy and strengthen the body.

| 加分配搭 Added benefits when paired with | |
| --- | --- |
| 白蘿蔔 Radish | 消食除脹 Aid digestion and relieve bloating |
| 山楂 Hawthorn (Shan Zha) | 祛斑消瘀 Remove pigmentation and stasis |
| 冬瓜 Winter melon | 開胃消食 Aid digestion |
| 芋頭 Yam | 健脾止瀉 Strengthen spleen and stop diarrhoea |
| 海帶 Seaweed | 祛濕止癢 Remove dampness and stop itch |
| 木耳 Fungus | 預防心血管病 Prevent cardiovascular disease |
| 番薯 Sweet potato | 降低膽固醇 Lower cholesterol |
| 白菜 Cabbage | 開胃消食 Aid digestion |
| 大蒜 Garlic | 加強維生素 B 族的吸收 Aid absorption of B Vitamins |
| 茄子 Eggplant | 增加血管彈性 Increase blood vessel elasticity |
| 香菇 Mushroom | 提高營養攝取 Increase absorption of nutrition |
| 豆苗 Peas shoot | 利尿消腫 Promote urination and relieve water retention |
| 竹筍 Bamboo shoot | 清熱化痰 Clear heat and reduce phlegm |

### 雞肉
### Chicken

▶保健功效 Health Benefits

具有益氣補精的功效。

Replenish vital energy and boost male potency.

| 加分配搭 Added benefits when paired with | |
| --- | --- |
| 枸杞 Wolfberries | 補五臟，益氣血 Nourish the organs, benefit vital energy and blood |
| 人參 Ginseng | 活血調經 nvigorate blood and regulate menstruation |
| 冬瓜 Winter melon | 益氣消腫 Replenish vital energy and relieve swelling |
| 海參 Sea cucumber | 益氣潤燥 Replenish vital energy, moisten dryness |
| 白蘿蔔 Radish | 使營養更全面 Increase nutritional value |
| 木耳 Fungus | 養氣補胃 Nourish vital energy and nourish the stomach |
| 紅豆 Red bean | 補血明目 Nourish blood, improve eyesight |
| 金針菇 Enoki mushrooms | 保肝健脾 Protect liver and strengthen the spleen |
| 包菜花 Cauliflower | 利五臟 Benefit organs |
| 洋蔥 Onion | 活血降脂 Invigorate blood, reduce blood lipid |
| 絲瓜 Sponge gourd | 清熱利腸 Clear heat and lubricate the intestines |
| 青椒 Green pepper | 預防動脈硬化 Prevent arteriosclerosis |
| 竹筍 Bamboo shoot | 提高營養攝取 Increase nutritional value |

### 鴨肉
### Duck

▶保健功效 Health Benefits

有滋陰養胃、利水消腫的作用。

Replenish the Yin, nourish stomach, promote diuresis and relieve

| 加分配搭 Added benefits when paired with | |
| --- | --- |
| 山藥 Chinese yam | 健脾養胃 Strengthen spleen and nourish the stomach |
| 酸菜 Pickled vegetables | 滋陰養胃 Replenish Yin and nourish the stomach |
| 白菜 Chinese cabbage | 降膽固醇 Lower cholesterol |
| 金銀花 Honeysuckle(Jin Yin Hua) | 滋潤肌膚 Moisturise the skin |
| 干貝 Dried scallops | 提高蛋白質 Boost protein intake |
| 豆豉 Fermented black beans | 降低脂肪 Lower fats |
| 芥菜 Leaf mustard | 滋陰潤肺 Replenish Yin and moisten the lungs |
| 冬菜乾 Pickled cabbage | 止咳潤肺 Stop cough and moisten the lungs |
| 薑 Ginger | 促進血液迴圈 Promote blood circulation |

現代人追求飲食健康，平日除了要吃得好，也要吃得對，哪些食材的配搭可以使營養、功效加分的呢？下列特別選出煲湯常用的肉類和蔬果，讓你平日在喝湯進補時，輕鬆讓健康加分！

Choosing the right food in your everyday meal is vital to a healthy well-balanced diet. Match the right ingredients to optimise their nutritional values. Use the below list of commonly used meat and vegetables for cooking soups for good health and best flavour.

## 烏雞 Black Bone Chicken

▶保健功效 Health Benefits

具有滋陰清熱、補肝益腎、健脾止血的作用。

Replenish the Yin, clear heat, nourish liver and kidneys, strengthen the spleen and stop bleeding.

| 加分配搭 Added benefits when paired with ||
|---|---|
| 紅豆 Red bean | 補血養顏 Nourish blood and improve the complexion |
| 竹笙 Bamboo fungus | 降低膽固醇的吸收 Reduce absorption of cholesterol |
| 蘋果 Apple | 解除疲勞 Reduce fatigue |
| 山藥 Chinese yam | 滋補身體 Nourish the body |
| 核桃 Walnut | 補鋅 Boost zinc intake |

## 黑豆 Black Bean

▶保健功效 Health Benefits

有活血祛風、消腫利水、補血安神、明目健脾、補腎益陰、解毒的功效。

Invigorate blood and dispel wind, promote diuresis and relieve swelling, nourish blood and calm the nerves, brighten vision and strengthen the spleen, benefit kidneys and Yin, detoxify the body.

| 加分配搭 Added benefits when paired with ||
|---|---|
| 穀類 Cereal and grains | 促進營養吸收 Promote absorption of nutrients |
| 紅糖 Brown sugar | 美容烏髮 Beautify complexion and darken hair |
| 番茄 Tomato | 促進血液迴圈 Promote blood circulation |
| 橘子 Orange | 促進維生素 C 的吸收 Aid absorption of Vitamin C |
| 柿子 Persimmon | 祛風解毒 Dispel wind and detoxify |

## 黃豆 Soy Bean

▶保健功效 Health Benefits

具有益氣養血、清熱解毒、美容養顏的功效。

Replenish vital energy, nourish blood, clear heat, detoxify and beautify the complexion.

| 加分配搭 Added benefits when paired with ||
|---|---|
| 花生 Peanut | 豐胸通乳 Good for breasts uplifting and firming, promote lactation |
| 紅棗 Red date | |
| 穀類 Grain and cereal | 促進蛋白質的吸收 Boost protein intake |
| 玉蜀黍 Corn | 促進消化 Aid digestion |
| 茄子 Eggplant | 養氣益血 Replenish vital energy, nourish blood |
| 香菜 Coriander | 祛風解毒 Dispel wind and detoxify |
| 牛蹄筋 Beef tendon | 防脊椎病 Prevent spinal disease |
| 白菜 Cabbage | 防乳腺癌 Prevent breast cancer |

## 白蘿蔔 Radish

▶保健功效 Health Benefits

能促進新陳代謝、增進食欲、清熱化痰、減肥的功用。

Facilitate metabolism, increase appetite, clear heat, reduce phlegm, promote weight loss.

| 加分配搭 Added benefits when paired with ||
|---|---|
| 雞肉 Chicken | 使營養更全面 Increase nutritional value |
| 豬肉 Pork | 健脾化痰 Strengthen spleen, reduce phlegm |
| 羊肉 Lamb | 溫補氣血 Warm and nourish vital energy and blood |
| 牛肉 Beef | 益氣補血 Replenish vital energy, nourish blood |
| 豆腐 Tofu | 促進消化 Aid digestion |
| 白菜 Cabbage | 潤膚養顏 Moisturize and beautify the skin |
| 蛤蜊 Clam | 強心護肝 Strengthen heart and liver |
| 金針菇 Enoki | 緩解消化不良 Aid digestion |
| 紫菜 Seaweed | 清肺熱，治咳 Clean lung heat, treat cough |

## 冬瓜
## Winter Melon

▶保健功效 Health Benefits

具有清熱解毒、利水消腫、減肥美容的功效。

Clear heat, detoxify, promote urination and relieve water retention, promote weight loss and beautify the complexion.

| 加分配搭 Added benefits when paired with ||
|---|---|
| 紅棗 Red date | 減肥潤燥<br>Promote weight loss, moisten dryness |
| 雞肉 Chicken | 益氣消腫<br>Replenish vital energy, reduce swelling |
| 鴨肉 Duck | 促進食欲，預防貧血<br>Improve appetite, prevent anemia |
| 火腿 Ham | 利尿 Diuretic |
| 蝦 Prawn | 幫助鈣質的吸收<br>Promote absorption of calcium |
| 蘆筍 Asparagus | 利尿 Diuretic |
| 蘑菇 Mushroom | 降血脂 Lower blood lipids |
| 海帶 Seaweed | 降血壓 Lower blood pressure |

## 蓮藕
## Lotus Root

▶保健功效 Health Benefits

具有滋陰養血、強筋健骨、補五臟之虛的功效。

Replenish the Yin, nourish blood, strengthen bones, tonify deficiency.

| 加分配搭 Added benefits when paired with ||
|---|---|
| 豬肉 Pork | 健脾養胃<br>Strengthen spleen, nourish stomach |
| 排骨 Ribs | 滋陰養血<br>Replenish the Yin, nourish blood |
| 糯米 Glutinous rice | 補中益氣<br>Replenish vital energy |
| 蓮子 Lotus seed | 補肺益氣 Nourish lung, replenish vital energy |
| 百合 Lily bulb (Bai He) | 潤肺止咳<br>Moisten lung, stop cough |
| 羊肉 Lamb | 潤肺補血 Moisten lung, nourish blood |
| 鴨肉 Duck | 清熱潤燥 Clear heat, moisten dryness |
| 生薑 Ginger | 清熱生津，止嘔吐<br>Clear heat, promote secretion of body fluids, stop nausea |
| 芹菜 Celery | 消除疲勞 Reduce fatigue |

## 銀耳
## White Fungus

▶保健功效 Health Benefits

具有強精補腎、潤腸益胃、滋陰潤肺、補腦提神、美容養顏的功效。

Boost male potency, nourish kidneys, moisten intestines, benefit the stomach, replenish the Yin, moisten lung, strengthen brain and invigorate the mind, beautify the complexion.

| 加分配搭 Added benefits when paired with ||
|---|---|
| 菊花 Chrysanthemum | 潤燥除煩<br>Moisten dryness, calm the nerves |
| 山藥 Chinese yam | 滋陰潤肺<br>Replenish the Yin, moisten the lung |
| 枸杞 Wolfberries | 美容養顏 Beautify the complexion |
| 百合 Lily bulb (Bai He) | 滋陰潤肺<br>Replenish the Yin, moisten the lung |
| 蓮子 Lotus seed | 減肥，祛斑<br>Promote weight loss, prevent pigmentation |
| 木耳 Black fungus | 潤肺補血<br>Moisten lung, nourish the blood |
| 魷魚 Squid | 防老，抗癌 Prevent aging, fight cancer |
| 蘋果 Apple | 止咳潤肺 Stop cough, moisten lung |

## 苦瓜
## Bitter Gourd

▶保健功效 Health Benefits

具有消暑除煩、明目解毒、養血益氣的功效。

Remove summer heat, calm the nerves, brighten the vision, detoxify, nourish blood and replenish vital energy.

| 加分配搭 Added benefits when paired with ||
|---|---|
| 瘦肉 Lean pork | 提供鐵質的吸收<br>Boost absorption of iron |
| 雞翅 Chicken wing | 健脾補胃<br>Strengthen spleen, nourish stomach |
| 玉蜀黍 Corn | 清熱解毒 Clear heat, detoxify |
| 豬肝 Pig liver | 補肝明目<br>Nourish liver, brighten the vision |
| 茄子 Eggplant | 抗衰老 Anti-aging |
| 洋蔥 Onion | 增強免疫力<br>Boost immunity |
| 雞蛋 Egg | 增強鈣質吸收<br>Boost absorption of calcium |

## 木耳
## Black Fungus

▶保健功效 Health Benefits
具有補氣活血、潤腸通便、滋潤強壯的功效。

Replenish vital energy, invigorate blood, moisten intestines to promote bowel movement, strengthen the body.

| 加分配搭 Added benefits when paired with ||
|---|---|
| 紅棗 Red date | 補血 Nourish blood |
| 銀耳 White fungus | 提高免疫 Boost immunity |
| 豬腰 Pig kidney | 提高免疫 Boost immunity |
| 大白菜 Chinese cabbage | 潤喉止咳 Moisten throat and stop cough |
| 綠豆 Green bean | 降壓消暑 Reduce stress and summer heat |
| 海帶 Seaweed | 降血脂 Lower blood lipids |
| 山楂 Hawthorn (Shan Zha) | 治潰瘍和經痛 Treat ulcer and menstrual pain |
| 雞蛋 Egg | 強健骨骼 Strengthen bones |
| 蘆薈 Aloe vera | 降低血糖 Lower blood sugar |
| 蝦 Prawn | 豐潤血色、毛髮<br>Give the skin a good colour, promote hair growth |
| 黃瓜 Cucumber | 減肥 Promote weight loss |
| 馬蹄 Water chestnut | 清熱化痰 Clear heat, reduce phlegm |
| 長豆 Long bean | 防高血壓、高血脂、糖尿病<br>Prevent high blood pressure, blood lipids and diabetes |

## 紅棗
## Red Date

▶保健功效 Health Benefits
具有補中益氣、養血生津、養肝防癌的功效。

Replenish vital energy, nourish blood, promote secretion of body fluids, tonify liver and prevent cancer.

| 加分配搭 Added benefits when paired with ||
|---|---|
| 桂圓 Dried longan | 養血安神 Nourish blood, calm the nerves |
| 黨參 Codonopsis (Dang Shen) | 健脾益氣 Strengthen spleen, replenish vital energy |
| 人參 Ginseng | 氣血雙補 Nourish vital energy and blood |
| 木耳 Black fungus | 補血調經 Nourish blood and regulate menstruation |
| 甘草 Licorice root | 補血潤燥 Nourish blood, moisten dryness |
| 黃豆 Soy bean | 豐胸通乳<br>Good for breasts uplifting and firming, promote lactation |
| 花生 Peanut | 補血 Nourish blood |
| 紅豆 Red bean | 健脾利水 Strengthen spleen, diuretic |
| 南瓜 Pumpkin | 補中益氣 Replenish vital energy |
| 糯米 Glutinous rice | 健脾養胃 Strengthen spleen, nourish stomach |
| 雞蛋 Egg | 益氣養血 Replenish vital energy, nourish blood |
| 白菜 Cabbage | 清熱潤燥 Clear heat, moisten dryness |

## 枸杞
## Wolfberries

▶保健功效 Health Benefits
具有補精益氣、保肝護腎、明目消渴、抗老降壓的功效。

Boost male potency, replenish vital energy, protect liver and kidneys, brighten vision, quench thirst, anti-aging and reduce stress.

| 加分配搭 Added benefits when paired with ||
|---|---|
| 菊花 Chrysanthemum | 養肝明目 Nourish liver, brighten the vision |
| 百合 Lily bulb | 補腎養血、寧心安神<br>Nourish kidneys and blood , calm the nerves |
| 豬肉 Pork | 滋陰補血 Replenish the Yin, nourish blood |
| 牛肉 Beef | 益精補血 Boost male potency, nourish blood |
| 雞肉 Chicken | 補五臟 Nourish the organs |
| 蓮子 Lotus seed | 養心益腎 Nourish the heart and kidneys |
| 鮑魚 Abalone | 養肝益腎 Tonify liver and kidneys |
| 蘋果 Apple | 和脾開胃 Balance the spleen and improve appetite |
| 竹筍 Bamboo shoot | 清肝降火 Clear the liver and lower heat |
| 銀耳 White fungus | 滋補健身 Tonify and strengthen the body |
| 蝦 Prawn | 補腎壯陽 Nourish the kidneys and boost male potency |
| 葡萄 Grape | 補血 Nourish the blood |

**功效**
**Effects**

這道湯具有益精補髓和清熱降火的功效。

This soup benefits spirits and bone marrow and relieves body heat.

# 菜乾豬肉湯
# Preserved Dried Vegetables with Pork Soup

## 材料

| | |
|---|---|
| 豬腱肉 | 150 克 |
| 豬龍骨 | 550 克 |
| 魷魚乾 | 3 只 |
| 白菜乾 | 150 克 |
| 紅棗 | 10 粒 |
| 蜜棗 | 3 粒 |
| 陳皮 | 1 小片 |
| 水 | 3.5 公升 |
| 鹽 | 適量（後下） |

## 做法

1. 將豬腱肉和豬龍骨洗淨剁塊，放入滾水中汆燙，撈起沖冷水，備用。

2. 魷魚乾放入滾水中汆燙，撈起沖冷水切條。

3. 白菜乾泡水，沖洗去沙，切段；陳皮洗淨，去白膜；紅棗和蜜棗略沖洗。

4. 將所有材料放入湯鍋中，以大火煮滾，轉小火煲約 3 小時，加入鹽調味即可。

## Ingredients

150g pork shank
550g pork bone
3 dried squids
150g preserved dried vegetables (Choi Kon)
10 red dates
3 honey dates
1 small piece dried mandarin peel
3.5 litres water
dash of salt (add later)

## Method

1. Rinse and cut pork shank and pork bone into pieces, blanch in boiling water, remove and rinse with cold water, set aside.

2. Blanch dried squids in boiling water, remove and cut into strips.

3. Pre-soaked preserved dried vegetables, rinse well to remove all sand particles, cut into sections. Rinse and remove dried mandarin peel white membranes. Briefly rinse red dates and honey dates.

4. Place all ingredients into a soup pot. Bring soup to a boil over high heat, and continue simmer over low heat for about 3 hours. Season with some salt to serve.

## Tips

- 因為白菜乾有很多沙子，所以一定要浸泡後才能使用。

- Preserved dried vegetables contain sand particles, so it is a must to pre-soak before cook.

▶ 常用海鮮乾貨 Commonly Used Dried Seafood

● 魷魚乾 Dried Squids
補血益氣
Replenishes blood, improves Qi

● 日月魚 Yat Yue Fish
補腎養肝
Tonifies the kidneys and liver

● 螺乾 Dried Conch
治心虛熱痛
Treats pain caused by heat within the heart

● 蠔乾 Dried Oyster
清肺滋陰
Clears lungs, nourishes the Yin

● 干貝 Dried Scallop
滋肝補腎
Nourishes the liver, tonifies the kidneys

● 鮑魚 Abalone
滋陰養顏
Nourishes Yin, maintains youth and beauty

功效
Effects

這道湯具有清熱解毒、瀉火利濕和生津止渴的功效。粉葛含有蛋白質、多種維生素和礦物質，具有清熱解暑的功效。

This soup relieves heat, detoxifies, benefits dampness in the body, improves saliva secretion and reduces thirst. Kudzu contains protein, various vitamins and minerals. It is effective in clearing body heat in summer.

# 粉葛蓮藕赤小豆煲魚湯
## Kudzu, Lotus Root, Azuki Bean with Fish Soup

材料

| | |
|---|---|
| 鮮魚 | 1 條（約 500 克） |
| 豬腱肉 | 400 克 |
| 粉葛 | 300 克 |
| 蓮藕 | 300 克 |
| 赤小豆 | 100 克 |
| 蜜棗 | 3 粒 |
| 干貝 | 2 粒 |
| 陳皮 | 1 小片 |
| 薑 | 2 片 |
| 水 | 3.5 公升 |
| 鹽 | 適量（後下） |
| 胡椒粉 | 適量（後下） |

做法

1. 魚處理乾淨，用少許油煎黃兩面，撈出放入煲湯袋，綁好。
2. 豬腱肉洗淨剁塊，放入滾水中汆燙，撈起沖冷水，備用。
3. 粉葛洗淨去皮，切塊；蓮藕洗淨，去皮切片。
4. 赤小豆、蜜棗和干貝略沖洗；陳皮洗淨，去白膜。
5. 將所有材料放入湯鍋中，以大火煮滾，轉小火煲約 3 小時。
6. 加入鹽和胡椒粉調味即可。

Ingredients

1 fresh fish (about 500g)
400g pork shank
300g kudzu root
300g lotus root
100g Azuki beans (small red bean)
3 honey dates
2 dried scallops
1 small piece dried mandarin peel
2 ginger slices
3.5 litres water
dash of salt (add later)
dash of pepper (add later)

Method

1. Rinse and clean the fish thoroughly. Pan-fry the fish with some oil until both side turn into golden brown. Transfer fish to a muslin bag, tie it up tightly.
2. Rinse and cut pork shank into pieces, blanch in boiling water, remove and rinse with cold water, set aside.
3. Rinse kudzu root, peel off skin and cut into pieces. Rinse lotus root, peel of skin and cut into pieces.
4. Briefly rinse Azuki beans, honey dates and dried scallop. Rinse and remove dried mandarin peel white membranes.
5. Place all ingredients into a soup pot. Bring soup to a boil over high heat, and continue simmer over low heat for about 3 hours.
6. Season with some salt and pepper to serve.

# Tips

- 若魚先煎過有去腥作用，煎過後可用熱水沖去表面油脂，煲出來的湯就會比較清澈不油膩。

- Pan-frying the fish will help to remove the fishy smell. Rinsing with hot water helps to remove excess oil, so that the soup will be clear and less oily.

功效
Effects

這道湯具有排毒、改善肌膚和清涼降火的功效。蘆薈也可修復細胞，並預防便祕的症狀。

This soup helps detoxify, healing and improves skin condition, and cools body heat. Aleo vera also helps repair skin cells and preventing constipation.

# 蘆薈紅棗木瓜湯
## Aloe Vera, Red Dates with Green Papaya Soup

### 材料

| | |
|---|---|
| 豬瘦肉 | 600 克 |
| 新鮮蘆薈 | 200 克 |
| 青木瓜 | 400 克 |
| 紅棗 | 15 粒 |
| 水 | 3 公升 |
| 鹽 | 適量（後下） |

### 做法

1. 豬瘦肉洗淨切大塊，放入滾水中汆燙，撈起沖冷水，備用。

2. 蘆薈去皮取肉，洗淨切大塊；青木瓜去皮除籽，切大塊；紅棗略沖洗。

3. 將所有材料放入湯鍋中，以大火煮滾，轉小火煲約 3 小時。

4. 加入鹽調味即可。

### Ingredients

600g pork lean meat
200g fresh aloe Vera
400g green papaya
15 red dates
3 litres water
dash of salt (add later)

### Method

1. Rinse and cut pork lean meat into big pieces, blanch in boiling water, remove and rinse with cold water, set aside.

2. Remove aloe Vera skin, rinse well and cut flesh into big pieces. Peel and remove green papaya seeds, cut into big pieces. Briefly rinse red dates.

3. Place all ingredients into a soup pot. Bring soup to a boil over high heat, and continue simmer over low heat for about 3 hours.

4. Season with some salt to serve.

### Tips

- 蘆薈表皮帶有黃色黏液，味苦而且有毒，吃了可能會引起肚瀉。所以皮要去乾淨，黏液也要洗淨。

- Aloe Vera skin contains yellowish slimy substance, it is taste bitter and contains toxins, consuming it may cause diarrhea. Therefore, it is important to completely remove the skin and rinse well the aloe Vera.

**功效**
**Effects**

這道湯具有改善血液和心臟、消除腸胃脹氣和養顏美膚的功效。佛手瓜含豐富膠原蛋白，可美肌潤膚，促進腸胃蠕動。

This soup improves blood and heart problems, removes flatulence and beautifies the skin. Chayote rich in collagen, improves and moisturizes the skin, and improves gastrointestinal motility.

# 佛手瓜銀耳雞爪湯
## Chayote, White Fungus with Chicken Feet Soup

材料

| | |
|---|---|
| 雞爪 | 12 支 |
| 雞胸肉 | 2 片（約 500 克） |
| 佛手瓜 | 2 個 |
| 銀耳 | 30 克 |
| 蜜棗 | 2 粒 |
| 水 | 3.5 公升 |
| 鹽 | 適量（後下） |

做法

1. 雞爪剪去爪尖，用鹽抓洗一下，沖洗乾淨再用麵粉搓洗一下，沖洗乾淨；雞胸肉去皮切大塊。
2. 雞爪和雞胸肉分別放入滾水中汆燙，撈起沖冷水，備用。
3. 銀耳泡水至脹發，洗淨去蒂；佛手瓜洗淨去皮去核，切大塊；蜜棗略沖洗，備用。
4. 將所有材料放入湯鍋中，以大火煮滾，轉小火煲約 3 小時。
5. 加入鹽調味即可。

Ingredients

12 chicken feet
2 pieces chicken breast meat
　(about 500g)
2 chayote
　(Buddha hand gourd)
30g white fungus
2 honey dates
3.5 litres water
dash of salt (add later)

Method

1. Cut off the chicken feet claws, rub with some salt, rinse well, rub again with plain flour, rinse again. Remove skin from chicken breast meat and cut into big pieces.
2. Blanch chicken feet and chicken breast meat in boiling water separately, remove and rinse with cold water, set aside.
3. Pre-soaked white fungus until expand in size, rinse and trim away the hard stem. Rinse chayote and remove skin and seeds, cut into large pieces. Briefly rinse honey dates, set aside.
4. Place all ingredients into a soup pot. Bring soup to a boil over high heat, and continue simmer over low heat for about 3 hours.
5. Season with some salt to serve.

## Tips

- 若不要雞爪和銀耳煲得太軟綿，中途可撈出雞爪和銀耳，待食用時再放回湯中一起食用。
- To avoid chicken feet and white fungus to become over soft, scoop out the chicken feet and white fungus once they are cooked and soft enough. Add into soup before serving.

這道湯能清熱潤燥，補中益氣的功效。
This soup excellent for nourishing, relieve heat and increases vitality.

# 霸王花豬骨湯
## Night Blooming Cereus with Pork Bone Soup

### 材料

| | |
|---|---|
| 豬龍骨 | 500 克 |
| 豬瘦肉 | 200 克 |
| 霸王花 | 100 克 |
| 香菇 | 3 朵 |
| 蜜棗 | 2 粒 |
| 薑 | 2 片 |
| 水 | 3.5 公升 |
| 鹽 | 適量（後下） |

### 做法

1. 將豬龍骨洗淨剁塊，放入滾水中汆燙，撈起沖冷水，備用。

2. 霸王花泡軟，洗淨切段；香菇泡軟，去蒂切半；蜜棗略沖洗。

3. 將所有材料放入湯鍋中，以大火煮滾，轉小火煲約 3 小時。

4. 加入鹽調味即可。

### Ingredients

500g pork bone
200g pork lean meat
100g night blooming cereus
　　　(Ba Wang Hua)
3 mushrooms
2 honey dates
2 ginger slices
3.5 litres water
dash of salt (add later)

### Method

1. Rinse and cut pork bone into pieces, blanch in boiling water, remove and rinse with cold water, set aside.

2. Soak night blooming cereus until soft, rinse well and cut into sections. Soak mushrooms until soft, remove stem and cut into halves. Briefly rinse honey dates.

3. Place all ingredients into a soup pot. Bring soup to a boil over high heat, and continue simmer over low heat for about 3 hours.

4. Season with some salt to serve.

### Tips

- 也可以使用大骨煲湯，先敲碎後再下湯，可讓骨髓的營養更容易釋放出來。
- 霸王花煲湯後清香甜滑，具有清心、健脾養胃及潤肺止咳的作用。
- May also used bigger pieces of pork bone. Crush the bones into smaller pieces to allow better release of nutrients.
- Night blooming cereus is fragrant and sweet when cooked in soups. It calms the mind, strengthens the spleen and stomach, moistens the lung and helps stop cough.

功效
Effects

這道湯具有清熱明目、止血涼血和改善眼熱赤痛的功效。
This soup relieves heat, brightens vision, stop bleeding, cooling blood, and help treat sore eyes.

# 蓮藕綠豆湯
# Lotus Root with Green Bean Soup

## 材料

| | |
|---|---|
| 豬龍骨 | 450 克 |
| 魷魚乾 | 3 只 |
| 蓮藕 | 500 克 |
| 綠豆 | 50 克 |
| 水 | 3.5 公升 |
| 鹽 | 適量（後下） |

## 做法

1. 豬龍骨洗淨剁塊，放入滾水汆燙，撈起沖冷水；魷魚乾放入滾水中汆燙，撈起沖冷水切條備用。

2. 蓮藕洗淨，去皮切厚片；綠豆洗淨。

3. 將所有材料放入湯鍋中，以大火煮滾，轉小火煲約 3 小時。

4. 加入鹽調味即可。

## Ingredients

450g pork bone
3 dried squids
500g lotus root
50g green beans
3.5 litres water
dash of salt (add later)

## Method

1. Rinse and cut pork bone into pieces, blanch in boiling water, remove and rinse with cold water. Blanch dried squids in boiling water, remove and cut into strips, set aside.

2. Rinse lotus root, remove skin and cut into thick pieces. Rinse green beans.

3. Place all ingredients into a soup pot. Bring soup to a boil over high heat, and continue simmer over low heat for about 3 hours.

4. Season with some salt to serve.

▶ 你知道嗎 Do You Know ?

綠豆的藥用價值非常高，綠豆加水煮至外皮脫落，濾除渣滓，直接取汁飲用。具有清熱解毒、調五臟、潤皮膚及潤喉止咳的功效。

Green bean has high medicinal value. May add some water and cook until outer skin off, strain the green bean water before use. It mainly clears heat, detoxifies, moderates the internal organs, moisturizes the skin and throat, and quenches thirst.

## Tips

- 蓮藕可以生吃，連皮榨成汁，清熱生津解渴，但脾胃寒虛者避免生吃，以防消化不良。

- Lotus root may be eaten raw, blend lotus root with skin on into juice. This is good for dispelling heat, promoting the production of fluids and quenching thirst. However, people with weak stomach should refrain from eating it raw as it is not easily digested.

**功效**
**Effects**

這道湯具有清熱降火的功效，是潤燥清熱的家常靚湯。西洋菜具有潤肺化痰，對咳嗽、咽乾舌燥及便祕有療效。

This soup helps to cool down body heat. An excellent soup for daily nourishing and relieve heat. Watercress helps to moisturize the lungs, reduces phlegm, and treats cough and dryness in throat and mouth, and constipation.

# 鴨胗西洋菜湯
# Watercress with Duck Gizzard Soup

材料

| | |
|---|---|
| 豬龍骨 | 400 克 |
| 豬腱肉 | 350 克 |
| 乾鴨胗 | 4 個 |
| 西洋菜 | 1 公斤 |
| 蜜棗 | 5 粒 |
| 南北杏 | 15 克 |
| 陳皮 | 1 小片 |
| 水 | 3 公升 |
| 鹽 | 適量（後下） |

做法

1. 將豬腱肉和豬龍骨洗淨剁塊，放入滾水中汆燙，撈起沖冷水，備用。

2. 西洋菜洗淨，泡入淡鹽水約半小時，撈出摘好，分嫩葉和梗部。

3. 鴨胗洗淨泡軟，每個切 3 片；陳皮洗淨，去白膜；其他藥材略沖洗。

4. 除了西洋菜嫩葉，將其他材料放入湯鍋中，以大火煮滾，轉小火煲約 2½ 小時。

5. 加入西洋菜嫩葉，小火煲約半小時。

6. 加入鹽調味即可。

## Ingredients

400g pork bone
350g pork shank
4 dried duck gizzards
1kg watercress
5 honey dates
15g sweet and bitter almond
1 small piece dried mandarin
   peel
3 litres water
dash of salt (add later)

## Method

1. Rinse and cut pork shank and pork bone into pieces, blanch in boiling water, remove and rinse with cold water, set aside.

2. Rinse well watercress and soak in salt water for about 30 minutes. Separate the young leaves and old stem of the watercress.

3. Rinse and soak dried duck gizzards until soft, cut each into 3 pieces. Rinse and remove dried mandarin peel white membranes. Briefly rinse all others herbs.

4. Except the watercress young leaves, place others ingredients into a soup pot. Bring soup to a boil over high heat, and continue simmer over low heat for about 2½ hours.

5. Add watercress young leaves, cook over low heat for about ½ hour.

6. Season with some salt to serve.

**Tips**

• 西洋菜洗淨後泡入淡鹽水中，可減少農藥，也可以去除匿藏的蟲子、水蛭等。

• Soaking watercress in salt water can help to remove chemicals and worms hidden in watercress.

**功效**
**Effects**

這道湯具有消逆氣和消食化積的功效。牛蒡具有利尿解熱，消腫解毒的功效。

This soup regulates flow of Qi, and aids in digestion. Burdock promotes diuresis, clears heat, reduces swelling and detoxifies.

# 白蘿蔔牛蒡湯
## Radish with Burdock Soup

材料

| | |
|---|---|
| 排骨 | 450 克 |
| 白蘿蔔 | 2 條（約 1 公斤） |
| 牛蒡 | 130 克 |
| 蠔乾 | 6 粒 |
| 干貝 | 6 粒 |
| 水 | 3 公升 |
| 鹽 | 適量（後下） |

做法

1. 排骨洗淨剁塊，放入滾水中汆燙，撈起沖冷水，備用。

2. 蠔乾和干貝略沖洗，浸泡備用。

3. 白蘿蔔洗淨，去皮切厚片；牛蒡洗淨，去薄皮切段。

4. 將所有材料放入湯鍋中，以大火煮滾，轉小火煲約 3 小時。

5. 加入鹽調味即可。

Ingredients

450g pork ribs
2 radishes (about 1kg)
130g burdock
6 dried oysters
6 dried scallops
3 litres water
dash of salt (add later)

Method

1. Rinse and cut pork ribs into pieces, blanch in boiling water, remove and rinse with cold water, set aside.

2. Briefly rinse dried oysters and dried scallops, and soak them separately.

3. Rinse radish, peel and cut into thick slices. Rinse burdock, peel slightly and cut into sections.

4. Place all ingredients into a soup pot. Bring soup to a boil over high heat, and continue simmer over low heat for about 3 hours.

5. Season with some salt to serve.

## Tips

- 牛蒡含鐵量高，削皮切段後可浸泡在水中，或在水中加幾滴醋，以防止氧化，但別泡太久。

- Burdock is rich in iron content, after peel and cut can soak in water, may also add a few drops of vinegar to the water to prevent burdock from oxidizing, but avoid soaking for too long.

功效
Effects

這道湯具有養肝補腎、強筋骨，緩解腰酸背痛，預防經期腰酸。

This soup helps nourish the liver and tonifies kidneys, strengthens muscles and bones, relieve backache, and prevent from menstrual backache.

# 杜仲玉竹補腰燉湯
# Du Zhong, Yu Zhu with Pork Bone Soup

材料

| | |
|---|---|
| 豬尾骨 | 600 克 |
| 杜仲 | 15 克 |
| 黃芪 | 15 克 |
| 玉竹 | 10 克 |
| 黨參 | 10 克 |
| 黃精（野山薑） | 10 克 |
| 斷續 | 10 克 |
| 當歸 | 5 克 |
| 紅棗 | 10 粒 |
| 熱水 | 1 公升 |
| 鹽 | 適量（後下） |

做法

1. 將豬尾骨洗淨剁塊，放入滾水中汆燙，撈起沖冷水，備用。

2. 所有藥材略沖洗。

3. 將所有材料放入燉盅，以中火燉約 3½ 小時。中途檢查外鍋的水，如果水太少就加熱水。

4. 加入鹽調味即可。

Ingredients

600g pork tail bone
15g eucommia bark (Du Zhong)
15g milkvetch root (Huang Qi)
10g Solomon's seal (Yu Zhu)
10g codonopsis
10g rhizome polygonti
　　(wild ginger)
10g dipsacus root (Duan Xu)
5g Chinese angelica
10 red dates
1 litre boiling water
dash of salt (add later)

Method

1. Rinse and cut pork tail bone into pieces, blanch in boiling water, remove and rinse with cold water, set aside.

2. Briefly rinse all the herbs.

3. Place all ingredients into a double-boiling pot. Double-boil over medium heat for 3½ hours. When there is insufficient water in the outer pot, top up with hot water.

4. Season with some salt to serve.

## Tips

• 女性食用這道湯可以預防經期腰酸。

• Women consume this soup can prevent from menstrual backache.

功效
Effects

這道湯可提供骨骼所需的膠質，並具有促進肌膚年輕、抗菌和增強免疫力的功效。海參具有補腎壯陽、益氣滋陰及通腸潤燥的作用。

This soup improves youthful skin, immunity and gives gelatin to the bones. Sea cucumber helps tonifies the kidneys, strengthens the Yang, improves Qi, nourishes the Yin, improves bowel movement and relives body dryness.

# 海參香菇燉雞爪湯
## Sea Cucumber, Mushroom with Chicken Feet Soup

材料

| | |
|---|---|
| 雞爪 | 10 只 |
| 雞胸肉 | 3 片（約 600 克） |
| 泡發海參 | 400 克 |
| 香菇 | 8 朵 |
| 小干貝 | 10 個 |
| 陳皮 | 1 小片 |
| 薑 | 2 片 |
| 熱水 | 1 公升 |
| 鹽 | 適量（後下） |
| 紹興酒 | 少許（後下） |

做法

1. 雞爪剪去爪尖，用鹽抓洗淨再用麵粉搓洗一下，沖洗乾淨；雞胸肉去皮切大塊。

2. 雞爪和雞胸肉分別放入滾水中汆燙，撈起沖冷水，備用。

3. 海參放入滾水中汆燙，撈起沖冷水，切厚片；香菇泡軟，去蒂切半。

4. 小干貝略沖洗；陳皮洗淨，去白膜。

5. 將所有材料放入燉盅，以中火燉約 3½ 小時。中途檢查外鍋的水，如果水太少就加熱水。

6. 加入鹽調味，也可加入紹興酒增香。

Ingredients

10 chicken feet
3 pieces chicken breast meat
(about 600g)
400g rehydrated sea cucumber
8 mushrooms
10 small dried scallops
1 small piece dried mandarin peel
2 ginger slices
1 litre boiling water
dash of salt (add later)
some Shao Xing wine (add later)

Method

1. Cut off the chicken feet claws, rub with some salt, rinse well, rub again with plain flour, rinse again. Remove skin from chicken breast meat and cut into big pieces.

2. Blanch chicken feet and chicken breast meat in boiling water separately, remove and rinse with cold water, set aside.

3. Blanch sea cucumber in boiling water, remove and rinse with cold water, cut into thick slices. Soak mushrooms until soft, remove stem and cut into halves.

4. Briefly rinse small dried scallops. Rinse and remove dried mandarin peel white membranes.

5. Place all ingredients into a double-boiling pot. Double-boil over medium heat for 3½ hours. When there is insufficient water in the outer pot, top up with hot water.

6. Season with some salt to serve. May add some Shao Xing wine for better fragrant.

## TIPS

• 木耳泡發海參在烹調前要反復沖泡、洗淨，然後放入鍋中，加入水、薑和蔥煮滾後，轉小火煲 20 分鐘即可。這樣就可以去掉海參的腥味。

• Rinse rehydrated sea cucumber repeatedly. Prepare a pot, add sea cucumber, water, ginger and spring onion, bring to a boil. Turn to low heat and continue simmer for about 20 minutes. This method will remove any odd smell from sea cucumber.

功效
Effects

這道湯具有補血、活血化瘀、治勞傷和祛斑養顏的功效。

This soup helps remedy for internal injury, improves blood circulation and removes spots to beautify skin.

# 田七干貝豬肉湯
## Tian Qi, Dried Scallop with Pork Soup

### 材料

| | |
|---|---|
| 豬瘦肉 | 600 克 |
| 黑木耳 | 15 克 |
| 干貝 | 4 粒 |
| 田七 | 10 克 |
| 紅棗 | 10 粒 |
| 蜜棗 | 3 粒 |
| 枸杞 | 5 克 |
| 水 | 3.5 公升 |
| 鹽 | 適量（後下）|

### 做法

1. 將豬瘦肉洗淨切塊，放入滾水中汆燙，撈起沖冷水，備用。

2. 黑木耳泡軟，洗淨去蒂；干貝略沖洗剝開；藥材略沖洗。

3. 將所有材料放入湯鍋中，以大火煮滾，轉小火煲約 3 小時。

4. 加入鹽調味即可。

### Ingredients

600g pork lean meat
15g black fungus
4 dried scallops
10g notoginseng (Tian Qi)
10 red dates
3 honey dates
5g Chinese wolfberry
3.5 litres water
dash of salt (add later)

### Method

1. Rinse and cut pork lean meat into big pieces, blanch in boiling water, remove and rinse with cold water, set aside.

2. Soak black fungus until soft, rinse and trim away the hard stem. Rinse and separate the dried scallops into pieces. Briefly rinse all the herbs.

3. Place all ingredients into a soup pot. Bring soup to a boil over high heat, and continue simmer over low heat for about 3 hours.

4. Season with some salt to serve.

### Tips

- 豬瘦肉脂肪少，是較健康的選擇。也可選用豬骨和豬腱肉，雖有少許脂肪，但香氣較多。豬肉經長時間煲煮後，脂肪會減少 30-50%。

- Pork lean meat contains less fat. Therefore, it is a healthier choice. Whereas pork bone and pork shank are contains more fat, but gives the soup more fragrance. However, the fats will reduce by 30-50% after boiling for some time.

▶ 你知道嗎 Do You Know ?

木耳
Cloud Ear Fungus

白背木耳
Hairy Wood Ear Fungus

燕耳
Swallow Fungus

皺木耳
Folds Fungus

秋木耳
Autumn Fungus

黑木耳具有清血脂的功效，可降低膽固醇、通血管，可預防中風和血管硬化，也可清內臟排毒。

Black fungus clears fat and heat in the blood, reduces cholesterol and clears the arteries. It can prevent stroke and vascular sclerosis and detoxify the internal organs.

功效
Effects

這道湯具有醒胃開胃、消食行滯、健脾益氣的功效。適合消化不良、胃口欠佳時飲用，可增進食欲。

This soup helps improves appetite aid digestion, strengthen spleen, and is very effective for those suffering from indigestion and loss of appetite.

# 鹹菜燒鴨湯
## Preserved Mustard Vegetable with Roasted Duck Soup

材料

| | |
|---|---|
| 燒鴨 | ½ 只 |
| 鹹菜 | 150 克 |
| 豆腐 | 2 塊 |
| 番茄 | 2 粒 |
| 鹹水梅 | 2 粒 |
| 薑 | 8 片 |
| 水 | 3.5 公升 |
| 鹽 | 適量（後下） |

做法

1. 將燒鴨剁塊，放入滾水中氽燙，撈起沖冷水，備用。

2. 鹹菜泡水，切大塊；番茄各切成 4 瓣；鹹水梅切半；豆腐切厚片，備用。

3. 除豆腐外，將所有材料放入湯鍋中，以大火煮滾，轉小火煲約 3 小時。

4. 加入豆腐煲煮片刻，嘗一嘗味道，若湯夠鹹則不需加鹽。

Ingredients

½ roasted duck
150g preserved salted vegetable
2 pieces tofu
2 tomatoes
2 salted plums
8 ginger slices
3.5 litres water
dash of salt (add later)

Method

1. Cut roasted duck into big pieces, blanch in boiling water, remove and rinse with cold water, set aside.

2. Soak preserved salted vegetables and cut into big pieces. Cut each tomato into 4 wedges. Cut salted plums into halves and cut tofu into thick slices, set aside.

3. Except tofu, place others ingredients into a soup pot. Bring soup to a boil over high heat, and continue simmer over low heat for about 3 hours.

4. Add tofu and simmer for just a while before serving. Taste the soup, if it's salty enough, no need add salt at all.

**Tips**

- 燒鴨要先氽燙去油，才能使湯水清澈不油膩；也可以把鴨皮去除才入鍋。

- Blanched roasted duck to avoid soup become too oily. May also remove the duck skin before using in the soup.

▶ 你知道嗎 Do You Know ?

潮州鹹菜
Teochew salted vegetables

榨菜 (Zha Cai)
Sichuan preserved turnip

福菜
Fu Cai

大頭菜
Turnip

酸鹹菜
Pickled mustard greens

梅菜 (Mui Choy)
Preserved mustard cabbage

鹹菜會使體內的鈉離子增加，風濕患者不宜多食用。
Preserved salted vegetables will increase sodium ions in the body, those suffering from rheumatism should avoid consuming it.

功效
Effects

這道湯具有增強心臟和補血的功能。消化不良、膽固醇高和痛風患者不宜食用。

This soup improves heart function and blood formation. People weak digestion, high cholesterol level and suffering from gouts should avoid consuming this soup.

# 豬心黑豆湯
## Pig Heart with Black Bean Soup

### 材料
| | |
|---|---|
| 豬心 | 1 個 |
| 豬龍骨 | 450 克 |
| 黑豆 | 1 碗 |
| 紅棗 | 10 粒 |
| 薑 | 2 片 |
| 水 | 3.5 公升 |
| 鹽 | 適量（後下） |

### 做法
1. 豬心洗淨，切成條狀，放入滾水中汆燙，撈起沖冷水；豬龍骨洗淨剁塊，放入滾水中汆燙，撈起沖冷水，備用。
2. 黑豆洗淨，以乾鍋炒至裂口；紅棗洗淨。
3. 將所有材料放入湯鍋中，以大火煮滾，轉小火煲約 3 小時。
4. 加入鹽調味即可。

### Ingredients
1 pig heart
450g pig bone
1 rice bowl black beans
10 red dates
2 ginger slices
3.5 litres water
dash of salt (add later)

### Method
1. Rinse and cut pig heart into strips, blanch in boiling water, remove and rinse with cold water. Rinse and cut pork bone into pieces, blanch in boiling water, remove and rinse with cold water, set aside.
2. Rinse black beans and dry-fry in a wok until crack open. Briefly rinse red dates.
3. Place all ingredients into a soup pot. Bring soup to a boil over high heat, and continue simmer over low heat for about 3 hours.
4. Season with some salt to serve.

### Tips
- 豬心買回來後用少許麵粉搓揉，放置約 1 小時，再用清水清洗，就可以去除異味。
- 豬心具有安神定驚、養心補心的功效，對失眠患者非常有幫助。

- Rub pig heart with some flour, and set aside for about 1 hour, rinse well to remove any stench from the meat.
- Pig heart has properties for calming the nerves. It nourishes the heart and is good for people suffering from insomnia.

功效
Effects

這道湯具有清熱潤肺、止咳化痰、生津止渴的功效。
This soup helps clear heats, moisturizes the lungs, relieves cough, reduces phlegm and quenches thirst.

# 花旗參海底椰螺乾湯
# American Ginseng, Sea Coconut with Conch Soup

## 材料

| | |
|---|---|
| 螺乾 | 50 克 |
| 豬瘦肉 | 250 克 |
| 乾海底椰 | 20 克 |
| 花旗參片 | 15 克 |
| 無花果 | 3 個 |
| 南北杏 | 10 克 |
| 薑 | 2 片 |
| 水 | 3.5 公升 |
| 鹽 | 適量（後下） |

## 做法

1. 螺乾洗淨泡軟，放入滾水中汆燙，撈起沖冷水；豬瘦肉洗淨切大塊，放入滾水中汆燙，撈起沖冷水，備用。

2. 所有藥材略沖洗。

3. 將所有材料放入湯鍋中，以大火煮滾，轉小火煲約 3 小時。

4. 加入鹽調味即可。

## Ingredients

50g dried conch
250g pork lean meat
20g dried sea coconut
15g American ginseng
3 dried figs
10g sweet and bitter almond
2 ginger slices
3.5 litres water
dash of salt (add later)

## Method

1. Rinse and soak dried conch until soft, blanch in boiling water, remove and rinse with cold water. Rinse and cut pork lean meat into big pieces, blanch in boiling water, remove and rinse with cold water, set aside.

2. Briefly rinse all the herbs.

3. Place all ingredients into a soup pot. Bring soup to a boil over high heat, and continue simmer over low heat for about 3 hours.

4. Season with some salt to serve.

### ▶ 人參的種類 Types of Panax Species

● 花旗參（西洋參）
American Ginseng
清火除煩
Clear heats

● 人參
Asian Ginseng
大補元氣
Replenishing the vital energy

## Tips

● 添加薑片可以去除螺乾的腥味。

● Add some ginger slices to remove any odour from conch.

● 參須
Panax Ginseng Root
生津止渴
Reduces phlegm and quenches thirst

● 紅參（高麗參）
Red Ginseng(Korean Ginseng)
補中益氣
Replenish the vital energy

# 四神湯
## Four-Sun Soup

材料

| | |
|---|---|
| 雞 | ½ 只（約 600 克） |
| 蓮子 | 20 克 |
| 芡實 | 15 克 |
| 淮山 | 15 克 |
| 茯苓 | 20 克 |
| 紅棗 | 6 粒 |
| 水 | 3.5 公升 |
| 鹽 | 適量（後下） |

做法

1. 雞去皮剁大塊，放入滾水中汆燙，撈起沖冷水，備用。

2. 所有藥材略沖洗。

3. 將所有材料放入湯鍋中，以大火煮滾，轉小火煲約 3 小時。

4. 加入鹽調味即可。

Ingredients

½ chicken (about 600g)
20g lotus seeds
15g Euryale seeds (Qian Shi)
15g Chinese wild yam
20g hoelen (Sclerotium Poriae Cocos)
6 red dates
3.5 litres water
dash of salt (add later)

Method

1. Remove skin and cut chicken into big pieces, blanch chicken in boiling water, remove and rinse with cold water, set aside.

2. Briefly rinse all the herbs.

3. Place all ingredients into a soup pot. Bring soup to a boil over high heat, and continue simmer over low heat for about 3 hours.

4. Season with some salt to serve.

這道湯具有促進食欲和減輕水腫症狀的功效。

This soup improves appetite and helps reduce water retention in the body.

# 何首烏烏雞湯
## He Shou Wu with Black Bone Chicken Soup

材料

| | |
|---|---|
| 烏雞 | 1 只（約 500 克） |
| 雞胸肉 | 1 片（約 200 克） |
| 何首烏 | 50 克 |
| 熟地 | 1 塊 |
| 黨參 | 15 克 |
| 紅棗 | 10 粒 |
| 水 | 3.5 公升 |
| 鹽 | 適量（後下） |

做法

1. 雞胸肉去皮切塊；烏雞斬成 4 大塊。
2. 雞肉放入滾水中汆燙，撈起沖冷水，備用。
3. 所有藥材略沖洗。
4. 將所有材料放入湯鍋中，以大火煮滾，轉小火煲約 3 小時。
5. 加入鹽調味即可。

Ingredients

1 black bone chicken (about 500g)
1 piece chicken breast meat (about 200g)
50g radix polygoni multiflori (He Shou Wu)
1 piece cooked rehmannia (radix rehmanniae)
15g codonopsis
10 red dates
3.5 litres water
dash of salt (add later)

Method

1. Remove chicken breast meat and cut into pieces. Cut black bone chicken into 4 parts.
2. Blanch chicken in boiling water, remove and rinse with cold water, set aside.
3. Briefly rinse all the herbs.
4. Place all ingredients into a soup pot. Bring soup to a boil over high heat, and continue simmer over low heat for about 3 hours.
5. Season with some salt to serve.

功效 Effects

這道湯具有生髮、烏髮和促進血液迴圈的功效。
This soup aids in hair-growth and promotes blood circulation.

**功效**
**Effects**
這道湯具有健脾利尿和清潤解暑的功效。

This soup strengthens spleen, diuretic and helps relieve thirst and heat.

# 節瓜眉豆煲魚湯
# Hairy Gourd, Black-Eyed Bean with Fish Soup

材料

| | |
|---|---|
| 鮮魚 | 1 條（500 克） |
| 豬龍骨 | 200 克 |
| 豬瘦肉 | 400 克 |
| 節瓜 | 500 克 |
| 眉豆 | 50 克 |
| 蜜棗 | 3 粒 |
| 陳皮 | 1 小片 |
| 薑 | 2 片 |
| 水 | 3.5 公升 |
| 鹽 | 適量（後下） |

做法

1. 魚處理乾淨，用少許油煎黃兩面，撈出放入煲湯袋，綁好。

2. 豬龍骨和瘦肉洗淨剁塊，放入滾水中汆燙，撈起沖冷水。

3. 節瓜洗淨去皮，切塊。

4. 眉豆和蜜棗略沖洗；陳皮洗淨，去白膜。

5. 將所有材料放入湯鍋中，以大火煮滾，轉小火煲約 3 小時，加入鹽調味即可。

Ingredients

1 fresh fish (about 500g)
200g pork bone
400g pork lean meat
500g hairy gourd
50g black-eyed beans
3 honey dates
1 small piece dried mandarin peel
2 ginger slices
3.5 litres water
dash of salt (add later)

Method

1. Clean the fish thoroughly. Pan-fry the fish until both side turn into golden brown. Transfer fish to a muslin bag, tie it up tightly.

2. Rinse and cut pork into pieces, blanch in boiling water, remove and rinse with cold water.

3. Rinse hairy gourd, peel off skin and cut into big pieces.

4. Briefly rinse black-eyed beans and honey dates. Rinse and remove dried mandarin peel white membranes.

5. Place all ingredients into a soup pot. Bring soup to a boil over high heat, and continue simmer over low heat for about 3 hours. Season with some salt to serve.

## Tips

● 節瓜可以後下，湯煲約 2 小時後再加入，這樣可以保持節瓜的口感。

● Hairy gourd can be added to the soup after about 2 hours of brewing. This will help to maintain the texture of the hairy gourd.

▶ 煲湯常用豆　Commonly Used Beans

● 眉豆 Black-Eyed Bean
健脾調中
Strengthens the spleen, moderates the torso

● 赤小豆 Azuki Bean
健脾胃
Strengthens the spleen and stomach

● 紅豆 Red Bean
補血消腫
Tonifies the blood, reduces swelling

● 綠豆 Green Bean
清熱解毒
Clears heat, detoxifies

● 黑豆 Black Bean
滋陰補血
Nourishes the Yin, tonifies the blood

● 黃豆 Soy Bean
益脾養中
Improves the spleen, nourishes the torso

功效
Effects

這道湯具有醒腦、增強呼吸系統和體力的功效。
This soup improves brain functions, breathing system and energy.

# 花旗參雞湯
## American Ginseng with Chicken Soup

### 材料
| | |
|---|---|
| 甘榜雞 | 1 只 |
| 雞胸肉 | 1 片（約 200 克） |
| 花旗參 | 12 克 |
| 枸杞 | 20 克 |
| 玉竹 | 20 克 |
| 淮山 | 25 克 |
| 茯苓 | 20 克 |
| 水 | 3.5 公升 |
| 鹽 | 適量（後下） |

### 做法
1. 雞胸肉去皮切大塊；甘榜雞去皮，斬成 4 大塊。
2. 雞肉放入滾水中汆燙，撈起沖冷水，備用。
3. 所有藥材略沖洗。
4. 將所有材料放入湯鍋中，以大火煮滾，轉小火煲約 3 小時。
5. 加入鹽調味即可。

### Ingredients
1 Kampong chicken
1 piece chicken breast meat
(about 200g)
12g American ginseng
20g Chinese wolfberries
20g solomon's seal (Yu Zhu)
25g Chinese wild yam
20g hoelen
(Sclerotium Poriae Cocos)
3.5 litres water
dash of salt (add later)

### Method
1. Remove skin from all chicken. Cut chicken breast meat into big pieces, and cut Kampong chicken into 4 parts.
2. Blanch chicken in boiling water, remove and rinse with cold water, set aside.
3. Briefly rinse all the herbs.
4. Place all ingredients into a soup pot. Bring soup to a boil over high heat, and continue simmer over low heat for about 3 hours.
5. Season with some salt to serve.

### Tips
- 花旗參不宜使用鐵制鍋具烹煮，而且服用花旗參時不宜食用白蘿蔔和茶。

- It is not advisable to use metal ware to cook American ginseng. Avoid having radish and tea when taking American ginseng.

▶雞的種類 Types of Chicken

- 白肉雞 Regular chicken
外形肥大，肉質鬆弛。
big in size with tender meat.

- 甘榜雞 Kampong chicken
低脂肪，久燉不爛。
low in fats and cooks well under prolonged cooking.

- 烏骨雞 Black bone chicken
低脂肪，高蛋白質，適合當藥膳。
low in fats and high in protein, suitable for making herbal soups.

**功效 Effects**

這道湯具有減輕氣喘症狀的功效。冬蟲夏草，簡稱蟲草，是傳統名貴滋補藥材，有助抗腫瘤、調節免疫、保護肝臟、呼吸系統和心血管系統等功效，是藥、食兩用之珍品。

This soup relieves shortness of breath and improves respiratory system. Cordyceps, also known simply as Chong Cao in Chinese, it is known to treat tumours, regulate the immune system as well as protect the liver, respiratory system and cardiovascular system. It is consumed both as medicine and food.

# 蟲草燉烏雞湯
# Cardycep with Black Bone Chicken soup

材料

| | |
|---|---|
| 烏雞 | 1 只（約 500 克） |
| 雞胸肉 | 1 片（約 200 克） |
| 冬蟲夏草 | 5 克 |
| 黨參 | 15 克 |
| 枸杞 | 5 克 |
| 熱水 | 1 公升 |
| 鹽 | 適量（後下） |

做法

1. 雞胸肉去皮切塊；烏雞斬成 4 大塊。

2. 雞肉放入滾水中汆燙，撈起沖冷水，備用。

3. 所有藥材略沖洗。

4. 將所有材料放入燉盅，以中火燉約 3½-4 小時。中途檢查外鍋的水，如果水太少就加熱水。

5. 加入鹽調味即可。

Ingredients

1 black bone chicken
  (about 500g)
1 piece chicken breast meat
  (about 200g)
5g cordyceps
15g codonopsis
5g Chinese wolfberries
1 litre boiling water
dash of salt (add later)

Method

1. Remove chicken breast meat and cut into pieces. Cut black bone chicken into 4 parts.

2. Blanch chicken in boiling water, remove and rinse with cold water, set aside.

3. Briefly rinse all the herbs.

4. Place all ingredients into a double-boiling pot. Double-boil over medium heat for 3½-4 hours. When there is insufficient water in the outer pot, top up with hot water.

5. Season with some salt to serve.

**Tips**

- 加一片雞胸肉可以讓湯水更濃郁一些。

- Adding a piece of chicken breast meat will increase the flavor of the soup.

▶你知道嗎 Do You Know ?

冬蟲夏草
Cordyceps

蟲草花
Cordycep flowers

北蟲草
North Cordycep

冬蟲夏草對人體有相當全面的保健功效，但價格昂貴；"蟲草花"並非冬蟲夏草的花，它是人工培養的蟲草子實體，培養基仿造天然蟲子所含養分，與常見的香菇、鮑魚菇等食用菌很相似，只是菌種、生存環境和生長條件不一樣。

Cordyceps (Dong Cong Xia Cao) is an expensive tonic with a wide range of medicinal benefits. Cordycep flowers are a cultivated fungus containing cordycepin and similar to shitake mushrooms and abalone mushrooms though with different growing conditions.

**功效**
**Effects**

這道湯具有改善呼吸系統、補氣和提升免疫力的功效。
This soup improves breathing, vitality and immune system.

# 黨參北芪雞湯
## Codonopsis, Astragalus with Chicken Soup

材料

| | |
|---|---|
| 甘榜雞 | 1 只 |
| 雞胸肉 | 1 片（約 200 克） |
| 北芪 | 15 克 |
| 黨參 | 30 克 |
| 淮山 | 20 克 |
| 茯苓 | 20 克 |
| 枸杞 | 20 克 |
| 龍眼乾 | 20 克 |
| 水 | 3.5 公升 |
| 鹽 | 適量（後下） |

做法

1. 雞胸肉去皮切大塊；甘榜雞去皮，斬成 4 大塊。
2. 雞肉放入滾水中汆燙，撈起沖冷水，備用。
3. 所有藥材略沖洗。
4. 將所有材料放入湯鍋中，以大火煮滾，轉小火煲約 3 小時。
5. 加入鹽調味即可。

## Ingredients

1 Kampong chicken
1 piece chicken breast meat
   (about 200g)
15g astragalus root
   (Radix Astragali)
30g codonopsis
20g Chinese wild yam
20g hoelen
   (Sclerotium Poriae Cocos)
20g Chinese wolfberries
20g dried longan
3.5 litres water
dash of salt (add later)

## Method

1. Remove skin from all chicken. Cut chicken breast meat into big pieces, and cut Kampong chicken into 4 parts.
2. Blanch chicken in boiling water, remove and rinse with cold water, set aside.
3. Briefly rinse all the herbs.
4. Place all ingredients into a soup pot. Bring soup to a boil over high heat, and continue simmer over low heat for about 3 hours.
5. Season with some salt to serve.

### Tips

- 疲倦乏力可以每星期都食用一次黨參北芪湯。
- To fight against fatigue, you may prepare and consume this soup once a week.

▶常用的保健藥材 Commonly Used Chinese Herbs

 ●淮山 Chinese Wild Yam
補益脾胃
Nourishes the spleen and stomach

 ●枸杞 Chinese Wolfberries
滋補肝腎
Nourishes the liver and kidney

 ●紅棗 Red Date
補氣健脾
Replenishes Qi, strengthens the spleen

 ●黨參 Codonopsis
補中益氣
Replenish the vital energy

 ●蓮子 Lotus Seed
養心安神
Eases the spirit, calms the nerves

 ●芡實 Euryale Seeds
健脾養胃
Strengthens the spleen, nourishes the stomach

功效
Effects

這道湯具有降膽固醇、高血壓和減輕水腫的功效。

This soup helps to lower cholesterol and high blood pressure, and reduces water retention in the body.

# 薏米海帶冬瓜豬骨湯
# Barley, Winter Melon with Pork Bone Soup

## 材料

| | |
|---|---|
| 豬龍骨 | 500 克 |
| 冬瓜 | 500 克 |
| 鮮海帶結 | 150 克 |
| 薏米 | 40 克 |
| 蜜棗 | 2 粒 |
| 薑 | 2 片 |
| 水 | 3 公升 |
| 鹽 | 適量（後下） |

## 做法

1. 將豬龍骨洗淨剁塊，放入滾水中汆燙，撈起沖冷水，備用。

2. 薏米浸泡約 1 小時，洗淨瀝乾；冬瓜洗淨，切大塊。

3. 海帶結和蜜棗洗淨。

4. 將所有材料放入湯鍋中，以大火煮滾，轉小火煲約 3 小時。

5. 加入鹽調味即可。

## Ingredients

500g pork bone
500g winter melon
150g fresh seaweed knots
40g barley
2 honey dates
2 ginger slices
3 litres water
dash of salt (add later)

## Method

1. Rinse and cut pork bone into pieces, blanch in boiling water, remove and rinse with cold water, set aside.

2. Soak barley in water for about 1 hour, rinse and drain well. Rinse and cut winter melon into big pieces.

3. Rinse seaweed knots and honey dates.

4. Place all ingredients into a soup pot. Bring soup to a boil over high heat, and continue simmer over low heat for about 3 hours.

5. Season with some salt to serve.

## Tips

- 冬瓜連皮一起煮，利尿消水腫的功效更佳。

- Cook the winter melon with the skin on as this has more effect of promoting diuresis and reducing water retention.

▶你知道嗎 Do You Know ?

海帶（昆布）
Kelp

海藻
Seaweed

群帶菜
Wakame

紫菜（海苔）
Dried Seaweed

海藻有助人體排毒、清血造血，促進代謝消化，清潔腸道，有效改善便祕。
Seaweed helps to detoxify our body, produce and purify blood, promote metabolism, and clear intestines of waste.

功效
Effects

這道湯具有清熱潤肺的功效。

This soup is good for treatment of ailments related to the lungs and throat.

# 無花果玉竹豬肉湯
# Dried Fig, Solomon's Seal with Pork Soup

材料

| | |
|---|---|
| 豬腱肉 | 450 克 |
| 豬龍骨 | 300 克 |
| 無花果 | 4 個 |
| 玉竹 | 25 克 |
| 淮山 | 15 克 |
| 紅蘿蔔 | 2 條 |
| 水 | 3.5 公升 |
| 鹽 | 適量（後下） |

做法

1. 將豬腱肉和豬龍骨洗淨剁塊，放入滾水中汆燙，撈起沖冷水，備用。

2. 紅蘿蔔去皮切大塊；無花果、玉竹和淮山略微沖洗。

3. 將所有材料放入湯鍋中，以大火煮滾，轉小火煲約 3 小時。

4. 加入鹽調味即可。

Ingredients

450g pork shank
300g pork bone
4 dried figs
25g Solomon's seal (Yu Zhu)
15g Chinese wild yam
2 carrots
3.5 litres water
dash of salt (add later)

Method

1. Rinse and cut pork shank and pork bone into pieces, blanch in boiling water, remove and rinse with cold water, set aside.

2. Remove skin and cut carrot into big pieces. Briefly rinse dried figs, Solomon's seal and Chinese wild yam.

3. Place all ingredients into a soup pot. Bring soup to a boil over high heat, and continue simmer over low heat for about 3 hours.

4. Season with some salt to serve.

## Tips

- 若不要豬腱肉煲得太碎散，中途可撈出熟軟的豬腱肉，待食用時再放回湯中一起食用。

- Over boiling may cause pork shank to become over soft and disintegrate into pieces. Therefore, scoop out the pork shank once they are cooked and soft enough. Add into soup before serving.

▶ 你知道嗎 Do You Know ?

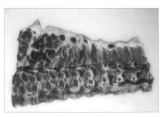
豬龍骨 Pork Bone

豬尾骨 Pork Tail Bone

豬大骨 Pork Leg Bone

豬龍骨含有大量的骨膠原，如果煲湯選用連豬尾的椎骨（豬尾骨），可以提高補精益髓的效果，也可以改善腰酸背痛，預防骨質疏鬆。

Pork bones contain a large amount of bone collagen, when using the tail bones with tail attached, the soup can boost male potency and strengthen the bones. It improves back aches and prevents osteoporosis.

功效
Effects

這道湯補而不燥，具有益氣養陰、清肺化痰和生津潤澤的功效。

This soup nourishes without drying, benefits Qi and Yin, clears the lungs, removes phlegm and aids in general body circulation.

# 海底椰雞湯
# Sea Coconut with Chicken Soup

## 材料

| | |
|---|---|
| 雞胸肉 | 2片（約500克） |
| 雞腳 | 10 支 |
| 新鮮海底椰 | 1包（約12個） |
| 百合 | 20 克 |
| 銀耳 | 15 克 |
| 蜜棗 | 3 粒 |
| 水 | 3公升 |
| 鹽 | 適量（後下） |

## 做法

1. 雞爪剪去爪尖，用鹽抓洗一下，沖洗乾淨再用麵粉搓洗一下，沖洗乾淨；雞胸肉去皮切大塊。
2. 雞爪和雞胸肉分別放入滾水中汆燙，撈起沖冷水，備用。
3. 海底椰洗淨，去皮切片；銀耳用水泡發，撕小朵；百合和蜜棗略沖洗。
4. 將所有材料放入湯鍋中，以大火煮滾，轉小火煲約 3 小時。
5. 加入鹽調味即可。

## Ingredients

2 pieces chicken breast meat (about 500g)
10 chicken feet
1 packet fresh sea coconut (about 12 pieces)
20g lily bulb
15g white fungus
3 honey dates
3 litres water
dash of salt (add later)

## Method

1. Cut off the chicken feet claws, rub with some salt, rinse well, rub again with plain flour, rinse again. Remove skin from chicken breast meat and cut into big pieces.
2. Blanch chicken feet and chicken breast meat in boiling water separately, remove and rinse with cold water, set aside.
3. Rinse sea coconut, remove skin and cut into slices. Soak white fungus until soft, tear in small pieces. Briefly rinse lily bulb and honey dates.
4. Place all ingredients into a soup pot. Bring soup to a boil over high heat, and continue simmer over low heat for about 3 hours.
5. Season with some salt to serve.

## Tips

- 雞肉去皮，湯水就不會太油膩。
- Remove chicken skin so that the soup would not be too oily.

▶ 你知道嗎 Do You Know ?

海底椰可潤肺補腎，而且養顏美容。
Sea coconut helps to moisturize the lungs, tonifies the kidneys, and also preserves youth and beauty.

海底椰乾
Dried Sea Coconut

新鮮海底椰
Fresh Sea Coconut

功效
Effects

這道湯具有清涼消暑、生津止渴的功效。椰汁和椰肉含大量蛋白質、果糖和維生素,有補脾益腎和催乳的功效。

This soup helps to cool down body heat and relieve thirst. Both coconut juice and flesh contain protein, fructose and vitamins. They can help tonify the spleen and kidneys and also help increase breastmilk supply.

# 老椰子竹蔗煲螺乾湯
## Coconut, Sugar Cane with Conch Soup

### 材料

| | |
|---|---|
| 豬龍骨 | 500 克 |
| 豬瘦肉 | 350 克 |
| 螺乾 | 50 克 |
| 老椰子 | 1 粒 |
| 竹蔗 | 250 克 |
| 蜜棗 | 2 粒 |
| 陳皮 | 1 小片 |
| 薑 | 2 片 |
| 水 | 3 公升 |
| 鹽 | 適量（後下） |

### 做法

1. 將豬瘦肉和豬龍骨洗淨切塊，放入滾水中汆燙，撈起沖冷水。
2. 螺乾洗淨泡軟，放入滾水中汆燙，撈起沖冷水。
3. 椰子剖開，取椰水過濾，備用；用刀子取出椰肉，切成條狀。
4. 竹蔗洗淨去皮，切段。
5. 陳皮洗淨，去白膜；蜜棗略沖洗。
6. 將所有材料放入湯鍋中，以大火煮滾，轉小火煲約 3 小時。
7. 食用前加入椰水煮片刻，加入鹽調味即可。

### Ingredients

500g pork bone
350g pork lean meat
50g dried conch
1 old coconut
250g bamboo cane
2 honey dates
1 small piece dried mandarin
  peel
2 ginger slices
3 litres water
dash of salt (add later)

### Method

1. Rinse and cut pork lean meat and pork bone in pieces, blanch in boiling water, remove and rinse with cold water.
2. Rinse and soak dried conch until soft, blanch in boiling water, remove and rinse with cold water.
3. Break coconut into halves, strain and retain the coconut juice for use later. Use a knife to remove coconut flesh and cut into strips.
4. Rinse bamboo cane, peel off skin and cut into sections.
5. Rinse and remove dried mandarin peel white membranes. Briefly rinse honey dates.
6. Place all ingredients into a soup pot. Bring soup to a boil over high heat, and continue simmer over low heat for about 3 hours.
7. Add coconut juice and cook for a while. Season with some salt to serve.

### Tips

- 椰水要最後加入，因為椰水久煮會失去香味和減低營養成分。
- Nutrients and fragrant in the coconut juice will be lost if it is heated for too long. Add the coconut juice before serving to prevent this from happening.

**功效**
**Effects**

這道湯具有祛濕利尿、降糖止渴、消暑清熱和清炎解毒的功效。

This soup can remove dampness, has diuretic effects, reduce sugar level, relieve thirst and heat.

# 苦瓜蠔乾黃豆排骨湯
## Bitter Gourd, Dried Oyster, Soybean with Pork Rib Soup

材料

| 排骨 | 500 克 |
|---|---|
| 苦瓜 | 1 條 |
| 小苦瓜 | 2 條 |
| 蠔乾 | 10 個 |
| 黃豆 | 100 克 |
| 薑 | 2 片 |
| 水 | 3 公升 |
| 鹽 | 適量（後下） |

做法

1. 將排骨洗淨剁塊，放入滾水中汆燙，撈起沖冷水，備用。
2. 兩種苦瓜切開，去籽和白膜，用鹽抓洗片刻，用水沖乾淨，切大塊。
3. 蠔乾大略沖洗，浸泡片刻備用。
4. 黃豆洗淨，以乾鍋炒香。
5. 除苦瓜外，將所有材料放入湯鍋中，以大火煮滾，轉小火煲約 2½ 小時。
6. 加入苦瓜，小火煲約 ½ 小時。
7. 加入鹽調味即可。

Ingredients

500g pork ribs
1 bitter gourd
2 small bitter gourds
   (dark green bitter gourd)
10 dried oysters
100g soybean
2 ginger slices
3 litres water
dash of salt (add later)

Method

1. Rinse and cut pork ribs into pieces, blanch in boiling water, remove and rinse with cold water, set aside.
2. Cut both types of bitter gourds, remove seeds and white portion, rub with salt, rinse well and cut into large pieces.
3. Rinse briefly dried oysters, soak for a while.
4. Rinse soybean and dry-fry until fragrant.
5. Except bitter gourd, place others ingredients into a soup pot. Bring soup to a boil over high heat, and continue simmer over low heat for about 2½ hours.
6. Add bitter gourd and cook over low heat for about ½ hour.
7. Season with some salt to serve.

**Tips**

- 小苦瓜更具清熱功效，加小苦瓜一起煲，清熱效果會更佳。

- The small bitter gourd is better for relieving body heat. Add small bitter gourd in the soup brings better effects to relieve the heat.

苦瓜
Bitter Gourd

小苦瓜
Small Bitter Gourd

**功效
Effects**

這道湯具有健脾利尿、預防心臟病、抗老和改善糖尿病的作用。
This soup strengthens spleen, diuretic, prevents heart diseases, anti-ageing and is suitable for diabetics.

# 紅蘿蔔玉米茶樹菇煲雞爪湯
# Carrot, Sweet Corn, Black Poplar Mushroom with Chicken Feet Soup

## 材料

| | |
|---|---|
| 雞爪 | 8 只 |
| 豬瘦肉 | 600 克 |
| 紅蘿蔔 | 2 根 |
| 玉蜀黍 | 1 根 |
| 茶樹菇 | 50 克 |
| 赤小豆 | 20 克 |
| 南北杏 | 15 克 |
| 陳皮 | 1 小片 |
| 薑 | 2 片 |
| 水 | 3.5 公升 |
| 鹽 | 適量（後下） |

## 做法

1. 雞爪剪去爪尖，用鹽抓洗一下，沖洗乾淨再用麵粉搓洗一下，沖洗乾淨；放入滾水中汆燙，撈起沖冷水。

2. 將豬瘦肉洗淨切塊，放入滾水中汆燙，撈起沖冷水，備用。

3. 紅蘿蔔洗淨去皮，切大塊；玉蜀黍去外葉留鬚，洗淨切段；茶樹菇泡軟，洗淨瀝乾。

4. 赤小豆和南北杏略沖洗；陳皮洗淨，去白膜。

5. 將所有材料放入湯鍋中，以大火煮滾，轉小火煲約 3 小時，加入鹽調味即可。

## Ingredients

8 chicken feet
600g pork lean meat
2 carrots
1 sweet corn
50g black poplar mushroom
20g Azuki bean
    (small red bean)
15g sweet and bitter almond
1 small piece dried mandarin
    peel
2 ginger slices
3.5 litres water
dash of salt (add later)

## Method

1. Cut off the chicken feet claws, rub with some salt, rinse well, rub again with plain flour, rinse again. Blanch in boiling water separately, remove and rinse with cold water.

2. Rinse and cut pork lean meat into big pieces, blanch in boiling water, remove and rinse with cold water, set aside.

3. Rinse carrot, peel and cut into big pieces. Remove sweet corn outer layer and retain corn silk, rinse and chop into sections. Soak black poplar mushroom until soft, rinse and drain well.

4. Briefly rinse Azuki bean and sweet and bitter almond. Rinse and remove dried mandarin peel white membranes.

5. Place all ingredients into a soup pot. Bring soup to a boil over high heat, and continue simmer over low heat for about 3 hours. Season with some salt to serve.

## Tips

- 玉蜀黍須是降血糖聖品，可以一起煲湯，別浪費。玉蜀黍若不是馬上使用，不要摘掉外葉，用保鮮膜包好，放入冰箱保存以免受潮長黴。

- Sweet corn silk helps to lower blood sugar, so do not waste them by throwing away. If sweet corns are not consumed immediately, store them properly by leaving the outer layer intact, wrap with cling wrap and place in the fridge.

# 沙葛素食藥材湯
# Vegetarian Herbal Soup

材料

| 沙葛 | 1 粒 |
| 枸杞 | 20 克 |
| 玉竹 | 20 克 |
| 洋參須 | 15 克 |
| 黨參 | 30 克 |
| 紅棗 | 15 粒 |
| 北芪 | 4 片 |
| 水 | 2.5 公升 |
| 鹽 | 適量（後下） |

做法

1. 沙葛洗淨，去皮切塊。

2. 所有藥材略沖洗。

3. 將所有材料放入湯鍋中，以大火煮滾，
   轉小火煲約 1½ 小時。

4. 加入鹽調味即可。

Ingredients

1 turnip (sengkuang)
20g Chinese wolfberries
20g solomon's seal (Yu Zhu)
15g American ginseng root
30g codonopsis
15 red dates
4 slices astragalus root (radix astragali)
2.5 litres water
dash of salt (add later)

Method

1. Rinse turnip, peel off skin and cut in big pieces.

2. Briefly rinse all the herbs.

3. Place all ingredients into a soup pot. Bring soup to a boil over high heat, and continue simmer over low heat for about 1½ hours.

4. Season with some salt to serve.

這道湯具有補氣、滋陰、解渴祛熱、提神和消除疲勞的功效。

This soup nourishes Qi and Yin, relieves thirst and heat, and removes fatigue.

# 五行蔬菜湯
## Five Elements Vegetable Soup

材料

| | |
|---|---|
| 沙葛 | 1 個 |
| 蓮藕 | 150 克 |
| 紅蘿蔔 | 1 條 |
| 高麗菜 | 1 顆 |
| 大頭菜 | 15 克 |
| 番茄 | 6 粒 |
| 黃豆 | 130 克 |
| 水 | 3 公升 |
| 鹽 | 適量（後下） |

做法

1. 大頭菜洗淨，浸泡約 15 分鐘，取出瀝乾。

2. 黃豆洗淨，以乾鍋炒香。

3. 全部蔬菜洗淨，沙葛和紅蘿蔔去皮切塊；蓮藕去皮切厚片；高麗菜切條；番茄切瓣。將所有材料放入湯鍋中，以大火煮滾，轉小火煲約 1½ 小時。

4. 加入鹽調味即可。

Ingredients

1 turnip (sengkuang)
150g lotus root
1 carrot
1 cabbage
15g preserved mustard (Tai Tow Choy)
6 tomatoes
130g soy beans
3 litres water
dash of salt (add later)

Method

1. Rinse and soak preserved mustard for about 15 minutes, remove and drain well.

2. Rinse soybean and dry-fry until fragrant.

3. Rinse all vegetables. Peel off turnip and carrot skin and cut into pieces. Peel off lotus root skin and cut into thick slices. Cut cabbage into strips and tomatoes into wedges. Place all ingredients into a soup pot. Bring soup to a boil over high heat, and continue simmer over low heat for about 1½ hours.

4. Season with some salt to serve.

功效
Effects

這道湯具有生津解暑、排毒利尿、降脂和活血補氣的功能。

This soup can relieve heat, detoxify, has diuretic effects, reduces fat, invigorate blood and nourish Qi.

# 老火湯

## 養生保健——從一碗湯開始

作　　　者　程安琪
攝　　　影　Shaun Sum

發　行　人　程安琪
總　策　畫　程顯灝
總　編　輯　呂增娣
主　　　編　李瓊絲、鍾若琦
編　　　輯　吳孟蓉、程郁庭、許雅眉、鄭婷尹
美 術 主 編　潘大智
執 行 美 編　李傳慧
美　　　編　劉旻旻、游騰緯、李怡君
行 銷 企 劃　謝儀方

發　行　部　侯莉莉
財　務　部　呂惠玲
印　　　務　許丁財
出　版　者　橘子文化事業有限公司

總　代　理　三友圖書有限公司
地　　　址　106 台北市安和路 2 段 213 號 4 樓
電　　　話　(02) 2377-4155
傳　　　真　(02) 2377-4355
E － m a i l　service@sanyau.com.tw
郵 政 劃 撥　05844889 三友圖書有限公司

總　經　銷　大和書報圖書股份有限公司
地　　　址　新北市新莊區五工五路 2 號
電　　　話　(02) 8990-2588
傳　　　真　(02) 2299-7900

製　　　版　興旺彩色印刷製版有限公司
印　　　刷　鴻海科技印刷股份有限公司
初　　　版　2014 年 11 月
定　　　價　新臺幣 188 元
Ｉ Ｓ Ｂ Ｎ　978-986-364-039-4（平裝）

國家圖書館出版品預行編目（CIP）資料

老火湯：養生保健：從一碗湯開始／程
安琪著 . -- 初版 . -- 臺北市：橘子文化，
2014.11
　面；　公分
ISBN 978-986-364-039-4( 平裝 )

1. 食譜 2. 養生 3. 湯

427.1　　　　　　　　　103022833

SAN YAU
http://www.ju-zi.com.tw
三友圖書
友直 友諒 友多聞

本書中所提供的食譜具有養生保健功效。有食療飲食都要注意適量或適性問題，有鑑於個人體質、年齡、性別或特殊情況而異，選擇具有調理性的食療，宜諮詢醫師的建議或診斷。